Abby B. Longstreet

Hospitality in Town and Country with Usages

Formal and Informal

Abby B. Longstreet

Hospitality in Town and Country with Usages
Formal and Informal

ISBN/EAN: 9783337163143

Printed in Europe, USA, Canada, Australia, Japan

Cover: Foto ©berggeist007 / pixelio.de

More available books at **www.hansebooks.com**

GOOD FORM

HOSPITALITY
IN
TOWN AND COUNTRY

WITH USAGES, FORMAL AND INFORMAL

*HOW TO MAKE IT A PLEASURE
TO ENTERTAINER AND ENTERTAINED*

By the Author of "Weddings, Formal and Informal"; "Cards,
their Significance and Proper Uses"; "Dinners"; "Manners, Good and Bad"; "Social Etiquette of New York," Etc.

NEW YORK
FREDERICK A. STOKES COMPANY
MDCCCXCII

TABLE OF CONTENTS.

HOSPITALITY

IN

TOWN AND COUNTRY

THE SOUL OF HOSPITALITY.

A PREMEDITATED hospitality in this country is an expression of fine civilization. At least it is meant to be.

Animals, in a natural state, share nothing eatable with their companions, until they are themselves surfeited. Educated human beings have learned, however, by experience, that prosperity is not complete until it is shared by others. It is only by dividing whatever material benefits we have that the soul of hospitality finds agreeable modes of expression.

5

As a rule, an unconsidered, ceremonious hospitality is a doubtful virtue. By unconsidered is meant a hospitality for which the comforts and diversions of guests, also its consequences upon the entertainer's family, are not duly considered.

True hospitality, as in other bestowals intended to be generous, has its highest impulse in a desire to confer pleasure, rather than in a wish to secure gratitude. It is not self-conscious.

Interested motives for extending hospitalities are pretty sure to be apparent. The entertainer may be convinced that his unrighteous secret is profoundly his own, but he deludes himself after the manner of those who wear wigs and use rouge.

Captious persons who are not, or perhaps cannot, be hospitable in large ways, assuage their pangs of envy by protesting against house-parties or large ceremonious dinners, saying that such entertainments are imported forms of sociability—as if good customs adopted from other people were a discredit to us.

Generous hospitalities are, by necessity, of recent date in America, because only lately has the growth of fortunes permitted large expenditures upon social pleasures. In England, France, and elsewhere, house-parties date back to feudal times, but in our country, except among patroons and Southern planters, they were rare—indeed, almost impossible half a century ago.

Even single guests, until within a few years, were

more likely to fix their own dates for visiting, and suit their own convenience in the length of their stay than they were to consult the pleasure of their entertainers. Hospitalities, under such circumstances, were not always an unmixed gratification either to guest or host, the latter's interest, as a rule, being an unconsidered matter. Hospitality was then a sacred duty, and sometimes a severe strain upon tolerance.

The host had one advantage in those times, if it be kind to call it an advantage, in that he was exempt from the duty of amusing his guest. Having assumed no social responsibility he was spared the fatigue of being diverting, except as friendliness suggested. Shelter and food were duly provided, and a welcome in manner, if not in words, was a solemn obligation, which no self-respecting person evaded, or thought of evading.

To amuse a guest is a recent custom. There may have been, and doubtless was, as fine a spirit in promptly and cheerfully giving a visitor what he wanted, just when he wanted it, as in informally inviting persons to be your guests when it is wholly at your convenience.

Our older style of bestowing hospitality was a dignified, though informal, mode of breaking bread with the needy. Its more recent usage is a bestowal of hospitality in expectation of an early equivalent in kind or contrast. It was not that the self-invited guest of former times was ordinarily needy in the

baldest sense of that word, but he really lacked something which might be only a variation in the monotony
of his life, and he sought what he thought he needed
under the roof of a friend or kinsman, and took it
without previous leave having been given. Indeed, it
was an especially formal courtesy among our not very
remote ancestors if notice of intention was given in
advance of a guest's arrival.

No householder declined such visits, except through
dire necessity, custom having made their acceptance
obligatory, and a realization—not then formulated in
speech—of a human brotherhood. There was a sentiment in such exchanges of courtesies that is sometimes
sadly missed from many of the most lavish entertainments of to-day.

Many American families that have become posess-
ors of large fortunes, from inherited social qualities,
indolence, or perhaps because they do not appreciate
their latent duty to share their abundance with their
fellows by entertaining them hospitably, live in a world
in which they are not even units, because they cannot,
or at least do not, recognize their obligations to civilization. They have not yet stirred to give themselves the
delight of being hospitable according to the best customs of their generation. These social drones fail to
get honey out of life because they do not know how
to gather it, and make no effort to find it in generous
human associations.

A perfect hostess is the most charming of women to

her friends, and the most loving and lovable one
in her own family. She is conscientious in the ex-
penditure of her resources, also of her leisure, which
is graciously and equitably divided between her guests
and her household. She knows that ostentation is an
evidence of inferior breeding, and that a frivolous
waste of time upon mere acquaintances is a wanton
sin, since those nearest and dearest to her have a right
to due proportions of it ; also, that the finest of mental
and moral cultivation is expressed by a just, and yet
gracious apportionment of all the good things with
which she is endowed.

Inequalities of fortune, when a moderate good fortune is unmistakable, are foolish reasons for missing
the pleasures of being hospitable. A host that spends
very much money too often feels less obligation to be
original in the diversions he provides. Monotony in
hospitality becomes very tiresome. To hesitate when
there is an opportunity to give pleasure, because of
necessity it must cost less money than Crœsus lately
spent upon his guests, is petty. It is witness to a
narrow spirit, since nobility of character and the
graces of accomplishments, lift a simple mode of being
hospitable up to the loftiest social levels—a height
that, with any amount of money, without such personal
allurements, the most ambitious of less well-bred men
and women cannot attain.

The most delightful of hosts are those who find out
the best qualities and attainments possessed by their

guests, and allure these graces out for mutual pleasure at a social gathering, either in town or country. Plain persons may be most gifted. If they are not personally charming, they usually cultivate and polish such talents as they have, but they are likely to be shy. Beauty is aware of the pleasure it provides simply by being, but the plain ones are adroitly called out by a clever host, and allured into expressing their best thoughts, or showing their special accomplishments. Perhaps they are made to pose unconsciously, that their few good points may be duly noted and properly recognized and valued. It is, as was said, the perfect hostess who acquaints herself with the special characteristics of all those whom she includes in her house-party, and she does not allow any of their special traits or qualities, that should be a distinguishing and unmistakable gratification to fellow guests, to be lost in the general brilliancy of her gathering.

One guest receives as much personal attention from hosts as another, unless age, uncommon acquirements, or a proper consideration for a stranger in the party, distinguishes her or him and entitles them to more than ordinary courtesies.

Sometimes a hostess discovers an unsuspected winsomeness or gift, in a plain, bashful woman or man, and, being its discoverer, sometimes is its amiable inventor —sweet angel that she is. She adroitly displays or explains what she has found to her other guests, and they too recognize an especial attraction, hitherto unknown.

If credulous, and not wholly blind to good parts that may have existed in secret in the least promising member of a happy house-party, they are able to make bright and handsome a dull countenance. "Believing is seeing," when a charming hostess points out the obscure fascinations of one of her guests to another.

A successful entertainer assorts her guests adroitly, so that there are no wounded vanities; and especially no self-respect is hurt. By tact she discovers who will be most sympathetic to each other, and which of them has that unnamable gift that compels others to appear all the better for being touched by a word, or called out by a question from them; and she carefully places such at her table and elsewhere accordingly.

A host or hostess is entitled to exemption from explanations or apologies to one guest for the presence of another, and she also has an indisputable right to resent the least lack of deference to any one of her party by one of her other guests.

There is no longer in the best American society that person who was once "placed before the rest."

However different the relations in life of two persons who have accepted invitations to be guests at one time, while they are in such mutual relations they are social equals. There may be other than social differences to be recognized, but these are duties of the host, which will be mentioned hereafter.

Wit that is neither biting nor too personal; attainments that are really worthy of regard and are the possession of those who are neither assuming nor presuming; friendliness that is not intrusive ; modesty that is not morbid nor silly; reclusiveness that is not exclusiveness and a kindliness that prefers another to itself, characterize the members of a perfect house-party of two or more guests.

The gifts or accomplishments of all guests are at the call of the entertainers, who have made themselves acquainted with the probable limits and availability of their entertaining traits.

Of course, in this picture of a social gathering, the too frequent " odd mate " in illy matched pairs must always be taken into account, and tactfully drawn into safe currents of conversation or diversion, which occasional necessity reminds one of George Meredith, when he writes, " The flame of the soul always rises upward. We must allow for atmospheric disturbances."

With unpliable, inharmonious members of a house-party—guests that could not be omitted—social atmospheric disturbances cannot always be avoided. The harmony of a host's best intentions are sometimes disturbed by elements that cannot be controlled, and it is then a real consolation to remember that the flame of the soul does rise upward, and that it is wisest and kindliest whenever possible to ignore the element that wrought discord.

Silence is said to be a social science when its application is kindly. With silence, censoriousness and mischievous gossip have no opportunity for thriving. Speech that is apt in time and tenor is akin to science. It heals wounded self-love, mends broken friendships, awakes the melody of jangled social notes. Nowhere is a gift and impulse of peacemaking more beautiful, nor anywhere does it win a higher appreciation, than at house-parties, made up of dissimilar temperaments and varying ambitions. This spirit of kindliness transforms a vexing tether that binds a group of persons together for a definite number of days, into a silken cord, the fastening of which is painful only when it is parted.

As a rule those who are unhappy at house-parties are such as measure and value each event and plan by its direct bearing upon themselves. If a guest expects to be wholly happy, his thoughts and motives must be to give, instead of striving to get, pleasure.

Hospitality is a mutual affair, if it be a success—an unmixed gratification. True happiness is unselfishness. Hospitality in thought and deed is its other name. The soul of the guest should be as large, fine and beautiful as that of the host, to make social perfection; and they are equal in the fineness of their intent when the broadest cultivation is expressed by the finest and most agreeable of manners.

Ideal friendships are born, nourished and established by the breaking of bread, and in those genial moods that follow bodily content, strive as one may to rise above this material fact. No one is able to escape it. If a person feels an unreasonable or an unjustifiable enmity toward another, he can find no easier reconciliation with him without the use of words, than by eating salt with him in his own house. Differences melt away in an atmosphere of hospitality.

TOWN VISITS.

THERE was a time, and not long ago, in America, when city houses were esteemed the most alluring places for prolonged hospitalities. With English examples, however, and larger capabilities for entertaining, a country house-party is almost equally attractive in winter and summer.

There are few private dwellings capacious enough in town, or perhaps few families that are small enough to make a large number of guests possible. Two visitors at one time is the most frequent number, although four, five or six are now and then gathered under one city roof. Four is an ideal group—that is, if wisely selected. It provides sufficiently diverse personalities, ages, gifts and attainments, to make time pass delightfully.

With a fortune, and an inclination to be generous in entertaining, without the experience of having been a guest at a house-party that was conducted in the most acceptable manner, or even having that theoretical knowledge that may be acquired from authentic sources, a hostess who fills her house for several days with guests, hazards her own peace of mind, and also that of her friends.

Nothing short of genius, or an amazingly good-fortune in her choice of comrades, spares the well-intentioned, ignorant hostess from making a pathetic failure of what ought to be a charming social event.

It is an exceptionally fine instinct that alone is capable of bringing persons, who were strangers to each other, into an atmosphere of serene cordiality and mutual sympathy for a number of consecutive days.

If my lady is an enthusiastic entertainer, and she knows little or nothing of the customary modes of amusing idle folk, in her effort to make her guests enjoy every hour of their stay with her, she is likely to be so energetic as to weary them by arranging too many diversions, not suspecting that such generosity is oppressive to such as may be familiar with seasonable recreations.

A hostess who is discreet will make herself aware of the characteristics of those whom she invites at the same time.

Each is supposed to have, in varying proportions, the natural graces and virtues of unselfishness, simplicity, modesty, sympathy, and above all, good-breeding. The latter is a necessity, and by this is not meant just a varnish or veneer of good manners that is not likely to wear off during a dinner or a reception, but which seldom endures the prolonged strain of a house-party.

Tact is an essential at every social gathering. It may be a talent, or it may be an acquirement, but it is an important gift for both host and guest. It is a high art, that need not be artificial; sometimes it amounts to genius, so delicate and influential is it. Whosoever has it, is *beau* or *belle*. Thus endowed, a woman need not have beauty, youth, a trained intellect, or a fortune.

Except in the case of near kinsfolk, when there are to be but two guests, they should be of the same sex, or there should be a wide diversity in their ages, the reason of which is apparent.

If a town house-party numbers more than two, it has not the significance of a pair, as it has where one is a young man, and one a young woman. Larger parties, of course, allow a freer choice, and sex need not be so carefully considered when making it up. Of course, if the host has grown sons or daughters at home, a party of two, one of each sex, is not in questionable taste, because thus the possible unpleasantness, or perhaps social indiscretion of what is called " Pairing guests by intention," is evaded.

A scheme for amusing a house-party is usually arranged in advance, certain of the plans of necessity being unalterably fixed. Others are, or may be, only suggested with alternatives. Sometimes a choice is given to visitors, but this is not always possible, since

most pleasant affairs require many days in which to
perfect them.

When there are several guests, the host selects
entertainments in which all may take part; therefore,
plans for them are ordinarily talked over not later than
the next day after arrival in town. At country house-
parties this custom is needless, as will be explained
further on. Suggestions for variations cannot be too
delicately made, and yet they may prove most helpful
to an entertainer, provided her strongest motive for
being hospitable is to give, rather than to receive,
pleasure.

It would be the worst possible taste to dissent, or
even to more than hint at alterations in a hostess's
intentions, or expectations, since guests have placed
themselves at her bidding by accepting her invitation.
An announcement of her plans is not obligatory,
many entertainers preferring, when they are definitely
acquainted with the tastes of their visitors, to make
each diversion a surprise. The latter plan cannot be
too deftly carried out, and sometimes only women
guests are consulted regarding divisions of time
between one pleasure and another. What they like
to do, or see that is not already arranged for, decides
plans, and it is a very ungallant man who dissents.

When there is but one guest, it is as kind to mention her own tastes and limitations to her hostess as
it is for the latter to cordially comply with her inclinations. Strangers to a town cannot always know who

or what is most interesting or most gratifying to their hosts, therefore it is bad form to insist upon anything. When guests have other friends or acquaintances residing in the same place, whom their entertainers do not know, or perhaps have not chosen to know, only a refined tact and a compliance with established usage spares each from giving offence.

A hostess may not disapprove of her guest's friend, but she may already have so extended an acquaintance that she is unwilling to enlarge her list, or there may be insurmountable social distinctions between them, that the guest does not, or perhaps cannot, perceive from her point of view, although the resident does all too clearly. Of course, the feelings and sympathies of each must be respected. To do this properly, the host suggests that her guest's card be sent with the date of a disengaged hour when the latter will be glad to receive her friends. This plan prevents interference with a hostess's arrangements; and she may or may not be present at the time such visits are to be made upon her guests. Of course, it is more kindly to see strange visitors, and such interviews create no further acquaintance beyond a casual recognition.

When there are no social affiliations between a guest's two sets of acquaintance in the same town, she cannot accept courtesies, or hospitalities from but one circle at a time, except at the suggestion of her host. Inequalities in position or attainments are being yearly more rigidly fixed in our youthful country that

more and more drifts astray from the spirit of a con-
stitution that was established on a presumed human
equality. Socially ambitious persons who have not
reached a desired position, and crave entrance into a
circle hitherto closed to them, either intentionally, or
inadvertently, sometimes try to make use of their
guest's, or their host's, acquaintance to advance their
purpose.

This is so adroitly managed that the instrument for
opening a way to introductions fails to suspect a purpose ; indeed is usually unconscious of ulterior motives
in apparently well-bred persons. Hence it is that
Good Form has been made to stand between what is
called " A pusher " and the " pushed." The latter do
not care to include the former in their visiting-lists, for
one reason or another, the customary one being that
already they have more acquaintances than they can
properly maintain. It is the pushing individual that
vulgarizes society in America, as well as elsewhere.

It would be diverting, if it were not pathetic, to
watch sordid-minded, clever men and women, while
making crafty uses of guileless hosts or guests, who
unconsciously are opening modes by which a coveted
society may be penetrated and captured. As a rule,
individuals with such offensive and dangerous talents,
ignore, or perhaps in the joy of a gratified ambition
they forget the hand that secured them admission to
what they wanted. Those who know the world most
thoroughly are not always the most delicate-minded

and most kindly of men and women; therefore it is that etiquette becomes a friendly protector that would, if it were strictly followed, save such as are in coveted circles from being wounded or imposed upon. Another name for etiquette is considerateness.

At large house-parties, breakfast is likely to be desultory; this plan meeting the ordinary host's convenience quite as well, or better than a prompt presence of everybody at once.

A hostess need not be present at breakfast if her children or other duties require her elsewhere—of course provided she has no young girl guests who are without chaperones, and there are men guests. If the hostess in the latter case cannot be present she arranges to send breakfast to the rooms of girl visitors.

This is of English usage, and is rigorously followed abroad. It has also become firmly rooted in good society here.

Even a young, unmarried daughter of the house, in the absence of her mother, does not descend to breakfast when there are men visitors, and her father is not present. When young girls are asked to be guests unaccompanied by an older woman, the hostess thereby assumes the office and responsibility of chaperone, and she more and more practises it, even if she does not feel its necessity. Of course, few American women do, but Good Form is Good Form, and it has excellent reasons for its establishment, though they may not be clearly apparent to such as have limited or insular social experiences.

When there is but one guest, or perhaps two or three, any lack of conformity to the customs of the family is unpardonable; especially is lateness to meals a proof of inconsiderateness, and a fine spirit is never thoughtless or indifferent to the convenience of others. Very much given to a selfish yielding to their own comfort, which is the least admirable variety of selfishness, are those who keep other persons waiting for them, either through a negligence or willingness to annoy others. To be late to breakfast, if the hostess has not distinctly explained that her guests are expected to have this in their rooms, or at table at their own time, is in especial bad form. This liberty is hardly possible without much exceptional care, unless there is a full corps of trained servants. Therefore considerate guests are likely to make as little disarrangement in the ordinary life of a hospitable family as is possible.

If what is called family worship is a custom, and the hour for its performance is mentioned to guests, their presence is obligatory. An absence from such service would be a serious discourtesy. It is proper to add that domestic devotions are less and less often practised in the presence of those not permanent members of the household. If the hour and place for them is not announced to visitors abiding in the house, or at least mentioned to them, their absence is expected.

To propose attending prayers without at least an implied permission would be an intrusion. A request for invitation to such sacred services should never be made. It would be an indelicacy, since each person's religion, and its mode of expression, is personal, and it ought to be allowed strict privacy, if preferred.

According to the ages and aptitudes of guests at a house-party are their diversions planned. Riding and driving, visiting museums and places of amusement, are the easiest of entertainments to provide, if the host is a person of fortune, which, of course, a house-party presupposes. Whatever is enjoyed by his guests is supplied by him. He purchases admission to whatever is to be seen or enjoyed, and he also provides horses for his guests, unless, as not infrequently happens, a woman has a favorite saddle-horse which she has been asked to send to her host's stable. Her groom accompanies the same, and may, if desired, bring his own animal.

This style of entertaining is, and must always be, for the few. It is mentioned here only because that few is growing in numbers and in luxury, and it is as common already in America as it is in England. Growing wealth is eagerly inquiring how inherited riches are and have been expending their fortunes, that they may do likewise, or perhaps better.

Persons with refined tastes, a moderate income and hospitable inclinations, who desire to give pleasure to one or two guests in a city home, can do it easily.

Their form of welcome is the same as that of very rich hosts, because certain matters, including friendship and its expressions, are always similar, but the luxury or lavishness of entertaining may be wanting.

Appreciative guests, however, fail to miss it, and they never fail to enter into the spirit and beauty of another's family life.

That elaborateness and munificence which characterizes house-parties of the very rich and very generous, is sometimes a compensation for similar pleasures received, with the expectation of returning them. Happily, this spirit is absent from simpler forms of entertaining. In the latter there is a more rational method of giving and getting enjoyment, because there are no especial changes made in the conduct of ordinary domestic life, and, of course, guests get nearer to the hearts of their entertainers. Friendships can take deeper root than in an artificial warmth.

If guests are fond of society, their coming has been announced by their host to such acquaintances and friends as will care to call, and perhaps also to invite them to visit, or at least, to breakfast in their homes. This is neighborly hospitality that has its habitual and mutual exchanges of courtesies to strangers that are staying in town.

It is a pleasant duty, indeed it is a religion with many of our most cultivated households, to offer some pleasant welcome to the guests of another. Not that they do this because it is a returnable politeness, but because they are sure it will be duly valued, and besides thus providing social gratifications for future guests of their own, they usually receive an immediate return in the different "points of view" that persons of other localities involuntarily give and receive when cultivated and broad-minded.

The most valuable and practical part of the education of some families has been acquired by entertaining wisely as to guests, and simply as to methods. That entertainer is a curious anomaly if he is both hospitable and *borné*. Of whatever is worth seeing or hearing, only the best and the choicest are offered to guests, when the spirit of the entertainer is fine. If the visit is to be brief, cards or seats for public places are secured in advance.

If it can be arranged, a little dinner or reception, or perhaps a luncheon, is given the day after the arrival of guests, care being taken to invite such friends as are most likely to be sympathetic with the guests' attainments or interests. They are impressively introduced, and allowed a few minutes' uninterrupted conversation. A hostess with discretion and tact is sure to find out a way to secure an opportunity for this preliminary to a pleasant acquaintance.

This chapter contains the heart of town hospitalities, but their formalities in detail are furnished under their own titles.

COUNTRY VISITS.

THE ideal place for entertaining large or small house-parties, or even single guests, is the country. Heaven and earth combine to rest the weary and to make idlers happy. Not nearly as much planning, originating, and providing amusements for guests are really required there, and yet many charming things may be done.

This is equally true of the country in summer and winter. One season has its shady nooks and lanes, its boating and flowers, and perhaps also its fishing and shooting; and the cold season has skating, tobogganing, sleighing and hunting. Both have riding and driving. The summer has warm moonlight nights, open verandas and song; and winter evenings have the firelight, music and dancing for the young, and storytelling, cards, and fine conversation for elderly hosts and guests, who furnish dignity and the graces of perfected society to house parties the year round.

More women than men are usually invited for extended visits in town, and doubtless this is due to the habits of men, who like clubs, and that freedom from an unfamiliar family routine which cannot be secured while house guests.

There are, however, always men enough within reach for dinners and dances when a house-party is made up of attractive persons.

In the country house-parties are likely to include more men than women, although this inequality is seldom intentional.

Liberty and open air pleasures delight most men, also most women, the latter becoming more enthusiastic as their skill in riding, boating and tennis increases, and their strength is proportionately improved.

The hostess, when she writes her invitations, mentions the names, and briefly, the tastes and acquirements of the guests she hopes to receive, in order that such as are not equally or similarly skilled, or perhaps are not in sympathy with what is mentioned as likely to be the usual amusements of the proposed house-party, may, if they choose, decline graciously, in order to allow others who are capable of contributing to the general enjoyment to be asked in their stead.

Not that each guest, man or woman, is expected to do everything, from music and dancing to athletics and rowing, but something each should be able to do, even if it be but story-telling, out-of-door and in-door photographing, fortune-telling by palms, playing dancing-music, or showing skill at cards. No one should be without a gift to contribute to the pleasure of a party; and every one is able to possess themselves with one or two of these agreeable accomplishments, unless he or she is willing to receive gratifications for which they make no return.

The host or hostess, each in their own province, informs guests soon after their arrival what there is to be done and seen, also what dinners and luncheons, etc., etc., are arranged. They mention when the mails arrive and depart, and where outgoing and incoming letters are placed; also the hours between which breakfast and luncheon may be eaten, except when these are to be formal, and to include others, invited from the neighborhood, of which ceremonial meals due announcements are made.

Dinner is always at a fixed hour, and guests are very much in need of tutoring in good manners if they are not assembled in the drawing-room a few minutes previous to its announcement.

Laggards at this important moment, or those who habitually keep others waiting, have need of valuable traits and fascinating characteristics to receive amiable tolerance for themselves.

Such bad-mannered persons need not go far to discover a justifiable reason for their being seldom invited to either large or small house-parties. Not that the most popular of hosts, or the most agreeable of guests, are martinets, but the loiterers are irritating, and spoil tempers, also dinners. They are never "thoroughbreds," because these fine-fibred individuals consider the comfort and the desires of everybody. Of course, pardonable delays fall to the lot of every one sometimes. It is only of the delinquents who are so often behind time that it is expected of them, that this severe criticism is made.

Personal talk about the temporarily absent is the most dangerous of ill-breeding at a house-party, gathered as its members often are from widely differing households, and sometimes from circles that touch each other only here and there. A possibility of falling into such unfortunate gossip is much more likely in town than in country. In the presence of nature such faults and foibles are apt to be forgotten, or left in the city atmosphere.

A host and hostess that are delicate minded, or perhaps only discreet, do not allow criticisms that hurt, or repetitions of scandals in their presence. A kindly word, or a charitable construction from them has spared many a listener from keenly wounded sensibilities at the mention of the misdeed, or suspected misdeed of kinsman or friend whose connection or association was unknown to intolerant or heedless chatterers.

Hosts are masters of the social situation, and with fine tempers and good hearts, they are able to make their qualities a happy contagion. If guests fall below their best standards of conduct, or lose a grip on their tempers, entertainers take the responsibility of setting them right, and they do it, with a generous and prompt firmness that those who give house-parties should not be without.

A host can always cool off a hot discussion, if he is himself cool, by begging the argument to be sus. pended—say until another season, or by some pleasant suggestion bring intolerant persons to a consciousness of their relations to hosts and guests, also to that spirit of hospitality which is a mutual obligation after each has assumed it by giving and taking bread and salt.

Unfortunately, too many persons forget that it is as hospitable to entertain another's opinions and thoughts for a little while, even if they cannot be adopted, as it is to give shelter and meat to each other. With this ideal of hospitality, hosts would entertain with more pleasure, and feel less apprehension of things going wrong; they would also experience less difficulty in placing dissenting persons in safe grooves of conversation, where, if they do not agree, they cannot overtly disagree and be offensive.

It is bad form to crowd too many diversions into one visit, and thus compel guests to hasten overmuch from one interest or amusement to another.

The most experienced and successful of entertainers are deliberate. They do not arrange for, much less insist upon having a ceremonious dinner-party after a picnic, nor a ball after an all day excursion. They have learned the charm of repose and the value of anticipation, both of which cannot be underestimated.

The young woman and young man who are most agreeable as visitors, are those who are ready to be useful, and who perceive little unsolicited ways for being agreeable. Assistance must be offered, as if it were a conferred favor to be helpful to a host, or to fellow-guests. After cordially proposing to be of service to their hosts, it would be an intrusion to press assistance upon them. Sincerity can always make itself understood. After the offer of assistance has been made, a prompt appearance at a time when there is a probability of being useful, is as far as a guest may prudently go.

There is always the shy, the physically delicate, or the one by temperament dull, and perhaps pessimistic, that the amiable visitor may draw out, cheer or encourage, and thus spare the hostess from suspecting them of being unhappy. With the pessimist, nothing depresses him so much as agreeing with him. As a rule he is not invited to house-parties, except he is also a humorist, and makes his despair picturesque, grotesque and diverting.

It may be his *metier* to pose as a hopeless being. If he does it artistically, he is not objectionable in a large party. In a small one he would be a " fee-faw-fum," and make everybody wretched.

If anybody has a grievance or a misery of any sort, and takes it into society, especially if society consists of a group of guests in a country house, he must conceal it—even permit no one to suspect that he has it. If it is a sorrow that is recognizable by his associates, it is in the worst of bad form to be one of a house-party, and risk depressing those who have an unquestionable right to uninterrupted gladness. To wipe one's eyes upon fellow-guests is unpardonable, and the worst of taste, leaving out the fact that it is a cruelty.

A smiling face is a reactive, and to practice cheerfulness while enduring a sorrow, makes sufferers more cheerful in reality.

This is one of Nature's own remedies; and Nature never makes mistakes in her method of healing.

IN America, women are the leaders in hospitality, because we have very few men with sufficient leisure. Therefore it is women who write all the invitations for visits and for house-parties that include husbands and wives, fathers and daughters, brothers and sisters, or women alone. Single men are invited, as a rule, by the man of the house, and very properly.

Of course, prolonged visits are usually considered some time in advance; although unanticipated ones, if made by but one or two, are frequent, even since more and more ceremonious laws are being enacted for visitor and visited.

Impromptu house-parties are occasionally arranged, especially in the winter, when there has been an abundant fall of snow, or a sharp frost, and there is much fine ice, that suggests them; but as such hasty groupings of persons while the social season is at its merriest is seldom possible, there is no etiquette established for them, except the strictest rules of chaperonage. This

33

is because dangers to health are forgotten by the young in inclement weather, and also the proprieties when frolicking is the motive for a house-party.

Good Form has adopted no exact usage for sending out invitations for ceremonious and fashionable house-parties, nor for house visiting generally, but regard for those whom an entertainer desires to receive, and who would like to accept, suggests as long an interval between the date of a request, and that of the visit, as is convenient. Five or six weeks is the usual time allowed between the invitation and the coming of visitors.

Sometimes invitations are given in the summer for winter visits, and sometimes they are planned a year in advance. The two forms last mentioned are, as a rule, verbal.

Invitations for prolonged visits are never in the third person, and, of course, are never engraved. At least they are not thus formal in America. A letter, or at least a cordial note, is always written to hoped-for house visitors—provided this request is not given in person. The day, also the hour for arriving is mentioned, also the length of the visit.

Etiquette is strict regarding such formalities. Those who accept hospitalities are as much benefited by this as those who bestow them. The hostess is thus able to made definite plans for her guest's pleasure, and she can also invite other guests to succeed such as depart early.

It is no longer in good form to accept any invitation that is general, and has no definite date. " Drop in at any time to dinner," or " Come and visit me," is an expression of comity and kindly welcome, and only that. It may mean a vague anticipation of pleasure if a visit could, some time or other, be arranged. Intimate indeed must be two friends who pass such complimentary expressions between each other, and really mean them with all that they imply. Of course, one may say, " Let me have a visit from you in the winter, or in the summer."

If this request has any meaning, later on an invitation will be duly sent, including its anticipated date. " Come to me for a week," if that is a convenient and agreeable length for entertaining a friend, or friends. For example :—" I'll be glad to see you on the afternoon train of the 10th of January." Then follows a list of the guests that are likely to be comrades, if any—also what is to be seen or done. The kindness of this explanation is apparent to women, who are thus made acquainted with the toilettes they will require. A man is always aware of his needs—a morning visiting suit, and an evening attire. His travelling clothes are his usual breakfast raiment, although he may appear in a frock coat, if he chooses.

Two days is the very longest time good manners allow between a reception of such invitations and answers. If replies are written at once, all the better, because more cordial, and because the hostess has more time for adjusting her plans, also in case of declination, to substitute other friends in the list.

Proper respect for a proposed hospitality would compel promptness in replying, even if convenience to the host did not demand it. If the hospitality includes several guests, it is called a "house-party," and all invitations are sent out at once, or as nearly at the same time as possible, and acceptances return near together. As each invitation includes the names of all the guests who are asked, and as acceptances and declinations are nearly at or about the same time, those who accept cannot know certainly who will decline, or who will be their fellow-guests, since others will be invited later, to make up a desired number. Of course, there are those whose replies are influenced by the names of others who are invited. Happily the entertainer is unable to give exact information at the date of her invitations, and it would be bad form to ask her later on, and in justice to her it is better that she is unable to write who are to be in her party.

Ordinarily, house-parties in town are invited for ten days. Sometimes replies mention that five days, or a week is as long as it is possible to have the pleasure of remaining with the proposed party. Should a guest accept for such a short period, the hostess may very properly, if she chooses, reply that, not being willing that her proposed coterie be broken up thus early, she will gladly invite such a guest at another time. She may, however, accept thus much time from her hoped-for guest, and ask another person to fill the place in the party.

There was a time when to be invited to conclude the visits of others, or to fill any vacated place, was not considered a compliment, but this sentiment has been reversed. Only dear and intimate friends are asked to take the chair that for some unforeseen occurrence is to be left unused at a dinner-party, and to be requested to take it is a marked compliment. Indeed, it is a favor that is not forgotten by a hostess. The same sentiment is expressed when an entertainer invites a friend to complete a broken house-party.

With a note inviting guests to a prolonged visit is enclosed a time-table for the route to be taken by them, with the train or steamer upon which they are expected duly marked. If it is difficult, or impossible to arrive at the hour indicated, a request for change of train is promptly made, and this note is at once answered with such explanation in detail as will allow a guest as much convenience and comfort as the altered plan permits.

Well-bred visitors will put themselves to the greatest inconvenience, when it is themselves or their hosts that are to suffer. Certainly they will take the train indicated, no matter how inconvenient it may be, if a change is to disarrange seriously a host's plans.

RECEPTION OF GUESTS IN TOWN.

IF women are to arrive alone by train in a town with which they are unfamiliar, it is proper that a member of the family meet them, or at least to send a trusty servant. Under any circumstances, it is a cordiality that, with many other pleasant details of hospitality, are being superseded by other forms that are more readily adapted to the changes of our social life.

Of course, public carriages are always awaiting incoming trains, and a prompt arrival at a host's residence is as much a certainty as any project well can be.

The fidgety woman who is in terror of a hackman is out of fashion. The dignified woman who is intelligent about her own expectations, prefers to go quietly to the residence of her friends. She has come to dislike greetings at crowded stations, where any excess of cordiality, or any lack of it, becomes matters of curiosity, if not of comment, to an idle crowd.

When men are visitors they are supposed to be able to take care of themselves always, but if strangers in a city, it is kinder to welcome them in person, and to accompany them to their destination. This attention, however, is not always possible when a party is to assemble on the same day, and at, or near, the same hour.

These explanations are made, not so much because the habit of meeting expected guests has been good form, as because it is falling into disuse, and likely to become obsolete in a short time. The conveniences for getting about a city have so largely increased within a brief time that meeting friends at stations is as unnecessary as it is to meet them on a train before their arrival at the terminus of their journey. It may be pleasant, but it is superfluous. Courtesy may be expressed better in other directions than by meeting incoming friends at the railways, or seeing them off, except by sending a servant, when guests are accompanied by none of their own.

Regarding the latter, Good Form is rigorous. If guests are not invited to bring a maid or a valet, they cannot come, except perhaps to stay long enough to unpack luggage and arrange it, and then perhaps return when the visit is terminated to repack it. In such cases at large houses there is usually a groom of the chambers, and as many maids as the party requires. Of course, this is a lately-established custom, and it could not be otherwise, because entertaining in large or lavish ways is also a very recent possibility.

Having been received by the hostess, and the host, if at leisure, it is considered a special courtesy if the hostess conducts a guest to her room after a brief interval for an interchange of friendly inquiries, during which a visitor's belongings are being deposited and unstrapped in her room. Married, or unmistakably older women at a house-party are always given the choicest rooms in America, where there are no inherited laws of precedence.

If the guest is a young lady, a daughter may take the hostess's place in showing her to the chamber assigned her. If arrivals are an hour or more previous to dinner or luncheon, a cup of tea or any other preferred light refreshment is sent in to the guests at once. Men visitors may be taken to their rooms by servants.

These small details may not appear to be very essential, and may seem unworthy of record to those unfamiliar with ceremonious entertaining ; but they are the fine finish of an elegant hospitality, and distinguish it from an unrefined entertainment. If those who are familiar with the perfection of entertaining pronounce these small directions unnecessary, they should consider those who have been less fortunate in their associations are more and more eager to acquire the polish of an older civilization, and it is agreeable to them to acquire it from written directions rather than by making an appeal to those who have long been in familiarity with it.

Guests are early informed of the dinner-hour, and how many are expected to be at table ; and then they

are left to repose until it is time to dress for luncheon
or for dinner, when, if the guest has no maid or valet,
assistance is proffered.

Of course, a meeting of hosts and guests usually
takes place in the drawing-room previous to the first
luncheon, and before dinner always, because at the
latter meal the etiquette of escorting to table and seat-
ing guests is likely to differ from day to day, accord
ing to the hostess's scheme for diverting her friends.
Usually the initial entertainment is a dinner-party that
includes persons from other houses whom she desires
to acquaint with the presence at her house of agreeable
visitors. Not infrequently a reception for them follows a ceremonious dinner.

Welcoming flowers are almost invariably placed in
guests' rooms to await their coming. This custom is
as common as it is charming. Good Form, also, and
delicate appreciation of the relations between hostess
and guests, requires that one or more of these blossoms
be worn at dinner. These are beautiful perfumed greetings from entertainer to entertained, and their silent
response at table is a fitting interchange of compli-
ments.

Anything and everything that refines and elevates
social life, is, or should be, obligatory upon the visitor
and the visited.

Guests' rooms should be supplied with all needful
materials for letter writing, including paper that is
stamped with the host's postoffice address, also seal-

ing-wax and tapers, and a calendar. It is a disputed question that etiquette refuses to be responsible for, whether or not postage-stamps should be included in the hospitable provision for correspondence, therefore each hostess is allowed to do as she chooses in this not very important matter.

Guests are informed by a card laid with the stationery when the letters are to be taken to the postoffice, and where to deposit them for the messengers; also, when letters may be expected, and where they will be placed when guests are not in the house to receive them. When there is to be an opera, a theatre, concert, or a drive very shortly after any meal, the same is mentioned in time for suitable dressing before descending.

Promptness in appearance in the drawing-room when the hostess expects guests is an obligation. Visitors, during the time of their stay in a house are, in a sense, part of their host's family; therefore, all persons who are invited during this time are formally presented to them. This custom is our own, and not in the least English, where introductions are by no means habitual, and by such omissions pleasant conversations are proportionately infrequent.

By receiving guests is not meant merely a welcome upon their arrival, but also presentations of others to them and of them to others. The quality of respect felt for guests is thus distinctly emphasized.

RECEPTION OF GUESTS IN THE COUNTRY.

IN the country, visitors are always met at trains or steamer landings, their invitations having informed them at what hour they are expected. This is usually a necessity, and is always an agreeable expression of welcome. Its informal formality is one of the many differences between town and country hospitalities. Even when the time comes that nobody goes to a city train to greet incoming friends, the country will keep to this cordial custom. Sometimes it may be only a carriage that awaits visitors, because circumstances compel a less complete welcome—but do not persons of rank in older countries send their carriages and servants to funerals to express their sympathy, and it may be, their sorrow? Luggage of visitors is cared for by orders of the host, and it is placed in guests' rooms while they are loitering for social reasons in their host's receiving-room.

As at town visits, a light refreshment is sent at once to a guest's chamber, if the hour is sufficiently

43

distant from the next meal, or it is late at night when
he arrives.

Flowers and everything pertaining to letter writing,
also all needful information regarding mails, with a
calendar, await guests. The hours between which
breakfast and luncheon are served, are mentioned and
the exact time for dinner, since the two earlier appointments for eating are usually "movable," when the hospitality is a house-party; but they are *en famille* when
there are but one or two guests.

Five minutes at least, and not more than fifteen, is
the usual time that ought to be spent in the parlor or
drawing-room previous to dinner, to allow for introductions, should there be new arrivals, or guests from
the neighborhood.

Evening toilettes, more or less ceremonious, are *de
rigueur*. Men must be in full dress, but the women
may wear what are called demi-toilettes, provided
there are no invited guests for dinner, and the number
of visitors is so small that it may not be considered a
house-party. Of course, circumstances regulate the
proprieties or fitness of a woman's costume for the
evening, but a man's is fixed for him. He cannot be
annoyed by the perplexities that trouble women's
souls regarding these things.

This matter is mentioned here, or rather it is re-
peated, because if invited guests shift the hours of
their arrivals from necessity, or for their own convenience, it is essential that they time their coming so as

to provide for making a toilette before dinner, etiquette in attire having become one of the tests of good manners.

Of course, circumstances may occur when dressing for dinner is impossible, and the host may beg his guest to forget it, and also his fellow-visitors, but since a fitting appearance is due to entertainers, as much as to self-respect, only an inevitable hindrance makes a lack of it pardonable. Usually women visitors arriving too late to dress for dinner appropriately, beg to be excused, and have dinner sent to their rooms. When there are other guests, especially if they are strangers, a considerate hostess makes this absence from a first dinner as tolerable as she can to her belated visitor. In appreciation of this kindness, the guest usually attires herself for evening, and comes down in time for tea and coffee in the drawing-room after dinner is over.

It is fine breeding not to mention the lateness of a new guest's arrival. Of course, if the belated guests have the characteristic of promptness, and their inopportune arrivals were due to an inevitable want of proper conveyance, in justice to themselves, and in recognition of a host's right to expect compliance with appointments mentioned in invitations, the cause of their failure to appear at the expected moment is briefly mentioned, and not dilated upon. Nothing is more tiresome to strangers than personal details that are important only in the mind of the one to whom they relate.

Introductions are made by host or hostess to all present when a guest arrives after dinner in the drawing-room, because they may be comrades for an extended time. If others have been present at the dinner, also, they are introduced because they were invited to meet the entire party, and not any particular member of it. This is an embarrassing formality sometimes, when guests are not of the social world, or if only lately. However, this is one of the penalties of being in what is called society, or, as usually happens, one of its pleasures, because it compels an entertainer to make them central figures, for at least a little while.

If a belated guest does not appear until morning, introductions are as general, but curiously enough, whether it is the absence of ceremonious attire, or simply the difference between merry morning and sedate evening, the process is much less grandly formal, and far more agreeable to those who are not ambitious of social distinctions.

The most charming of all visits is made singly in the country, either by a man or woman. There is a special fascination in receiving and being received with a cordiality that is undivided, and a sincerity that means personal friendliness, and not an aggregated social function. A guest who is to be the sole visitor is always met by a member of the host's family, and he or she enters the real household at once. In this way one makes acquaintances with the children and friends immediately, and becomes familiar at once with all the interests of a country home ; also all the sweet and simple delights that such a visit implies.   If not an invalid, or in need of special repose, the break-fast hour, that sweetest of all the day, is spent with the entire family.

ENTERTAINMENTS FOR GUESTS IN TOWN.

WHEN an invitation reads, for example, "Come on Monday, the 8th of December, for ten days, and have a quiet time with us—just us," and adds "An afternoon train arrives from your place at 3 : 30," a sincere hostess means what she has written. If she finds her guest is in a mood for amusement—and most guests are—her preferences are delicately consulted, so that she may not feel that she is requesting, much less dictating, pleasures that may not be equally acceptable to the one entertained. It is pleasant to gratify the tastes of guests, and it is a dull hostess who cannot find out what these are without direct questionings.

Nothing is more vulgar than to discuss, in their presence, the cost of diversions that guests are to share ; nor is a refined visitor inquisitive about this, or indeed about anything that is not openly discussed in a household, because surprises for a visitor are sometimes the pleasure of an entertainer. When she diverts guests in that manner she must be very sure of them, also of herself, and failures to give and receive pleasure are imminent.

It is safest to proffer amusements and outings, rather than to urge them. To provide an atmosphere of sunny freedom is the finest and best of all hospitalities, when entertaining, and especially when there are but one or two guests.

At house-parties there must be one who controls and leads and who plans and executes, of course, with the enthusiastic co-operation of guests; but when a hostess has only a special friend, or perhaps two or three friends, she may give dinners, luncheons, an afternoon or evening reception, go upon excursions with her guests, or take them to see whatever is worthy of their attention or within the compass of their interests, tastes or sympathies.

It is quite as much the duty of a guest to enter heartily into the purposes of a hostess as it is the hostess' dutiful satisfaction to confer pleasure. Any inadvertence of a hostess who may have many cares besides that of being an entertainer, should be passed unnoticed, and always is by her fine-fibred, considerate visitors, while the religion of hospitality requires that a hostess be not too exacting in a matter of moods on the part of her guests.

Temperament is the friend or enemy of individuals, and it should be taken into consideration by those who are thrown together.

Observant persons are convinced that we are to be

more pitied or envied for our temperaments than for any ill or good fortunes. Neither hosts nor guests can be too much impressed by this physiological fact, because it too often happens that the most admiring of friends really do not know each other until they have passed days together, since when meeting at other times, all have been in their best spirits. The other side of a sweet disposition, the dissatisfied and dark side, may never have been suspected. Doubtless owing to this human variation of spirits was coined the proverb that " To really know anybody we must summer and winter with them."

The same kind of entertainments are provided for house-parties that are given to a smaller number of guests, but more of them are crowded into the same time. Dinners, luncheons, germans, and less formal dances, musicals, little house plays, for which a skeleton of events is filled out with impromptu conversations and dramatic actions, etc., etc., with sufficient intervals for a proper amount of recuperation, are arranged for youthful parties. All these things are put in train before the arrivals of guests, who, as a rule, assemble within a day or two of each other and at the same date when it is possible.

Of course, all the guests that are available are asked to take parts in whatever amusements are proposed, and no matter how distasteful a character may be, guests are compelled by good-breeding to assume the one the host assigns to them. It is said, and doubtless it is true, that the least alluring character in a play as it is read, may by study be made the most artistic one in a little drama. Even Bottom, with an ass's head, in " Midsummer-Night's Dream," has made himself now and then a first star in the comedy.

With consciousness of a guest's latent talent or aptitudes, directors in little house diversions may assign undesired parts to certain guests who are unaware of their own abilities in such lines. Those who are captious about the parts assigned them, as a rule, testify to their vanity or their incapacity.

ENTERTAINMENTS FOR GUESTS IN THE COUNTRY.

OUT-OF-DOOR amusements at country houses must necessarily depend very largely upon their geographical situation.

Under roofs amusements resemble each other, no matter where they are provided. The most essential differences there are between diversions arranged for guests in one house and another, or between one season and another in the country, depend upon the number of guests. A large party may do what a small one does; but a small one is limited by its size to certain diversions. It is a foregone conclusion that hosts who invite large parties have the means at command for amusing them in an acceptable manner, indoors and out, of course provided their guests are able to co-operate in their schemes.

For single guests, or for two or three, there are drives and rides through lanes and woods, midday walks in shady places, where luncheon may await them amid rocks, as if provided by the Brownies ; calls upon country neighbors, who are all charmed by such invasions, provide tea, etc., and perhaps display interesting houses and gardens, after which they may find dinner awaiting them while the sun is vanishing, set out upon veranda tables, or in the house, and feel a sweet and rare remoteness from all the cark and care and the grinding fatigues of city living. The closing of a fair day, as Keats has it, when " The sun lays his chin on the green wood weary, with all his poppies gathered round him," is a dream, but it is a dream of realities that makes life worth living.

These are but a few of the simple and satisfying pleasures which the "sole and only," or a pair of guests, may enjoy at a country house with genial hosts. Of course, the evening has music, fine conversations, reading aloud, story-telling, cards, and perhaps a wood fire, if the weather be damp, or there is any other justification for the delightful companionship of a blaze upon hospitable hearths.

Of course neighbors and near-by friends are invited to make acquaintance with a guest or guests, at dinner or breakfast ; the latter being one of the most delightful of gatherings at a country house. Curds and cream, fresh fruits and field flowers provide an ideal morning banquet that lingers in a city dweller's memory among other poetic recollections when costly banquets are all forgotten.

Those who are invited to meet a neighbor's guests

return the courtesy by entertaining them in turn. This is etiquette that is based upon the Golden Rule, since permanent dwellers amid beautiful simplicities have no comprehension of the charm their surroundings have for city folk—who for the most part of their lives are sandwiched in between bricks—unless they learn to appreciate these delights by contrast, when they make return visits to a crowded town.

For companies large enough to be called house-parties, there are rides and drives, guests sometimes being invited to bring their own horses and grooms when distance is not a formidable objection to this arrangement, and stables of hosts are ample. There may be picnics or surprise luncheons in picturesque places; dances, private theatricals, musicals, *bals poudrés,* or fancy dress parties, to which outsiders may be invited; dinners of ceremony with guests from other houses; luncheons, formal and informal, the first only when others than the house-party are present, and the latter a movable feast, which is eaten as each one chooses, from twelve-thirty o'clock to two-thirty; tennis, billiards, racket, bowling; also boating and sailing; and, at proper seasons, shooting, hunting and fishing, the last-mentioned diversions depending upon the location of the house. Except shooting, none of these diversions belong exclusively to men, some women being experts in each.

Programmes for in-door and out-of-door amusements are made out in advance, those outside being subject to the weather, but those in-door are inalterable as to date ; therefore, the guests are expected to be in readiness to fill their parts or to carry out their plans duly and promptly. Chaperones and elderly persons are invited to take active parts in amusements, but it is no offence if they decline, while, except in case of indisposition, no young persons may refuse to do whatever is asked of them by their entertainers. They may perhaps be consulted regarding a part they can have in a play, but as a rule they are not, because a freedom to select would too often overturn what might have been a charming success.

If a character in a performance is not equal to what they believe to be their *metier*, guests can prove their right to a better one by doing well that assigned them. The usual objection made to a soubrette part in a play, is "I shall look horrid."

Very well, look horrid. The contrast when restored to the ordinary appearance will be more impressive than before. The "Audrey" of Shakespeare has not infrequently been the fascination of a performance.

To be late at rehearsals, to object to a manager's arrangements, or for a manager to be unwilling an actor should do original work with an old character that is hedged around with traditions, is a serious mistake ; indeed it is bad form, especially as it is for amusement, and not to make fame or fortune, that a play is enacted at a house-party in the country.

At a *bal poudré*, costumes may be simple or elegant,

historic or original, but as the name of the amusement suggests, they are usually historic; the hair being powdered white, gold, copper or silver, but oftenest it is white.

A host cannot be too careful when mounting his guests, if he does not know their skill in the saddle, or whether rough roads are familiar to them; also if they know how to treat an animal wisely, and knowing how, will do it.

At large house-parties the hostess, and sometimes the host, does not appear at breakfast. The former may have children with whom she chooses to spend a part of each day, and the host may have his estate, his "correspondence," etc., etc., to look after. In such cases an unchaperoned girl who is under the especial care of the hostess, asks that her breakfast be sent to her room—if she is properly conventional. If in her heart she is rebellious against her limitations, all the same she remains away from the party during her hostess's absence, because she respects the office, and perhaps the sentiments of her hostess, and will not add to her cares, as an entertainer. With the young girl's acquiescence a hostess may transfer her duty to a woman guest that has sufficient age and dignity to fill the position of chaperone during the early part of the day, and badly-bred and indelicate is the young woman guest who is not deferential to her delegated care taker.

It is customary at many house-parties, where there

is one young girl guest under the special protection of the hostess, for her to attend to the flowers for the table, and elsewhere. This occupation, except there is an excursion to which she may or may not go, according as she wills, and according to what older woman there is in the outgoing party, occupies much of her morning most agreeably. She and the gardener are in consultation, and they arrange charming schemes of color or fragrance, which is properly appreciated, and duly admired. Even though there is a person employed to adorn the house daily with flowers, the young lady who assumes this care is likely to produce effects that are less formal and more charming than a florist can, unless he is an artist in his occupation.

A hostess always appears at luncheon if possible, and this is a merry meeting for everybody. It is always a substantial meal, with cold meats, one or two vegetables, bread, butter, fruit and melons, also chocolate and tea. Wine is seldom given at luncheon, but sparkling or still waters, in two or more qualities, are always provided.

The latest magazines, and most recent books and games are among the entertainer's gratifying assistants in making hospitality gracious, eloquent and easy. They allow her a leisure that nothing else can, or does, because if she is absent from her guests, and she has not furnished them with diversions in the way of books, papers, new music, etc., she is apt to be worried lest they are suffering *ennui*.

However many and capable her servants may be, making time pass agreeably with her guests when there are no active or special diversions in progress, is a duty for which others are useless. She cannot transfer the office of entertainer to anybody.

To let them alone a part of every day is usually a kindness to them, but there are guests and guests, and some of them are exceedingly difficult to amuse, these usually believing that constant diversions are their due.

Of course this sort of visitor is always considered a duty guest, and is invited with that dreary thought in mind. If hosts could enter a family with invitations, with liberty to take and to leave whomsoever they chose, their house-parties might be an unadulterated pleasure; but unhappily they cannot. It is fortunate for families that are glad to make their difficult member even tolerably contented that they cannot. There is, as a rule, neither sufficient pity for, nor enough effort made to give pleasure to those who are miserable by nature, and who do not know how to be happy or glad about anything for more than an hour at a time.

This sort of guest is, perhaps, the most dissatisfying element in plans for entertaining either large or small parties.

Alas, that the arid plains of life should be too largely populated with persons born with discontent as their special and unavoidable inheritance! It is use-

less to insist that they might improve their temperaments if they wished, since they were not endowed
with a desire to be other than they are.

The popular house visitor is he or she who rec-
ognizes this fellow-guest's misfortune, and tries to
make it as tolerable as possible for them, also for
their hosts.

The question was asked one day of one of the most
delightful of entertainers how she happened to invite
so disagreeable a visitor, and she replied sweetly but
gravely, "Because nobody else seemed willing to, and
the poor soul has always herself to be miserable with."

Excessive expenditures for the table at country
parties is as vulgar as it is everywhere. It never con-
vinces anybody that hosts who are inordinately lavish
were bred to abundance.

Those who feel contempt for a reasonable frugality,
also for a generosity that is conscious of a to-morrow,
in money matters, are those who have not possessed it
long enough to give it its proper place in their respect.
Hosts who are hospitable in the most acceptable and
beautiful ways, were, as a rule, born to fortunes, and
know how to expend them. The newly rich who are
able to be properly reserved in their generosities,
especially at their tables in the country—yes, and in
town also—must have a genius for it.

There may be cases of atavism, but to whatever, or
whoever such hosts owe it, they are fortunate to know
by instinct how to draw a fixed line between gener-

osity and lavishness, and to maintain it gracefully and graciously.

. Extravagant outlays of money, or a continuous devotion to the entertainment of many people, never delude those who benefit by these, nor gain their respect. A world whose regard is worth winning, long ago determined that fine breeding in giving and receiving hospitality, was expressed by a proportionate diversion between time, money and family seclusion.

LEAVE-TAKING AND DEPARTURES IN TOWN.

To overstay a time mentioned in an invitation is bad manners. To remain longer than the host at first requested must be as an especial and unmistakable favor to an entertainer, and never to suit the convenience or a desire on the part of a guest.

Nor is it good form in a host to press for an extension of the time first accepted. Even the most gratifying of visits are pleasanter to remember when terminated reluctantly by visitors and visited. The most desirable of guests usually have pre-arranged disposals of their leisure, and it would be unpleasant to be urged to add to an allotted time, should a hostess so far forget herself as to beg for it.

Indeed it is a species of egotism that urges visitors to remain when they have planned to take leave. It is the same characteristic that allows persons to say of others, "They are anxious that I should visit them," as if anxiety was possible in such a matter.

Perhaps they selected an unconsidered word, but it sounds pompous to those who use a language properly.

A hostess, if she be a perfect entertainer, is sure to say she is sorry a visit has terminated, and she adds a hope for another when it can be arranged. Sometimes, as we said earlier, parties are arranged a year from date, but unless the same group is desired, the invitation is given in private.

Most guests never permit a disarrangement of family orderliness when taking leave of their hosts, however much the latter may press them to have an early breakfast, luncheon or dinner. All trains and steamers provide food for travellers, and to put a host to the least inconvenience is unpardonably bad form. Indeed most hosts have already fallen into a custom made easy by the recent luxuries of travelling, of giving themselves or their servants no care for the appetite of outgoing guests, unless they are aged, or are invalids. For the latter kind of guests, all of the old-fashioned impulses of tender, nourishing attentions are keenly alive, and practically followed.

When a man is to depart early in the morning from a town house, he takes leave of his entertainers the night before, and a cup of coffee or tea, and a wafer, are taken to his room in the morning by a servant to whom has been assigned the duty of wakening him at the proper moment. Sometimes a very considerate, energetic young man prefers to spend the night at a hotel after making adieus with his hostess, and doubtless this custom will become a usage, as it very properly should. His luggage, except that in his hands, may always be sent to a city station the night before, and properly checked to its destination.

The same is true of a woman's belongings, if she is compelled to make an early morning departure, and they cannot be taken upon a carriage that conveys her to a train or steamer. Of course with a woman guest who has no attendant of her own, a servant is sent to see to her luggage, if it has not been checked the night before, and at departures at any time of the day, the same goes to carry her light parcels and to deposit them with her when she has taken her place for her journey. Whether it is late or early in the day that a departing guest leaves her host's residence, it is not a rigid custom that compels any member of the family to see her off. If it is a sad leave-taking, neither those who go or those who stay are willing to risk displaying emotion before strangers, therefore sentiment combines with the luxury and safety of travelling to eliminate the old habit of accompanying—sometimes by whole families and groups of families—an outgoing friend at a pier or station.

When a woman, youthful or elderly, is to make a journey alone, and it is to continue several hours, or days, it is a kindly and usual custom to send a few dainty refreshments with her. These are packed in little boxes or baskets, with napkins, that are kept for the purpose of exchanges in travelling hospitalities.

Not that good food is not procurable on trains and boats, but a hostess's thoughtfulness makes hers more agreeable than any which can be purchased, and it takes the place of the old-fashioned floral good-by. Sometimes a bouquet is tied to the luncheon case, but it is small and significant, and not the former monster which suggested that its bearer was on her way to a horticultural fair as a competitor.

Our later form of "speeding the parting guest," appears to have reached perfection. At least its improved methods are in accord with our advanced facilities for travel and for being hospitable. Distances made partings and greetings of vital importance but a little while ago, but miles have ceased to make separations, and the sadness, and joy, also the dignity, of coming and going, has fallen away from among our sensibilities. For this reason among others, our ceremonies with incoming and outgoing visitors have changed, and very properly. Our present formalities, or perhaps, our informalities, compare with our former customs very much as do our social usages with those of an Oriental.

LEAVE-TAKING AND DEPARTURES IN THE COUNTRY.

IN the country, as in town, visitors who are familiar with the best social usages and practice them, never overstay the time for which they were invited, nor do hosts who are equally well acquainted with the best customs, urge them to extend their stay. This is due to no lack of hospitality, or cordiality, but because it is unkind to disarrange the carefully considered plans of either guest or host. A delicate entertainer claims to be the one who is obliged by the visit of a friend, and therefore it would be selfish to beg for more of her guest's company.

A more practical reason for concluding a visit at the allotted time, is, that hosts may have planned for the arrival of another group of guests, or have engagements for themselves.

It is unpleasant to be urged against one's own desires; hence, and for reasons already mentioned and for others, a visit must conclude under all ordinary conditions as the invitation for it arranged that it should.

Unless it is an unavoidable necessity, guests do not depart by very early morning trains. Considerateness for a host and his servants at a country house make such departures in bad form. All avoidable disarrangements of a host's customs are bad form.

Under one pretext or another guests should remain until it is a convenient hour in the day for their host to get them to a station, even though this postponement of their departure inconveniences them, a fact which is usually known to their entertainer. Of course, a host with the true spirit of hospitality desires to make the close of a visit even more charming than its commencement, and he is sure to plan to do it, but the thoroughbred man and woman refuses to allow a household to be disarranged when a little personal sacrifice might prevent it.

If guests cannot avoid going away from their entertainers before the usual time for breakfast, leave-taking, and all proper courtesies are completed the night before, and coffee or tea is taken to their rooms by a servant who has been ordered to waken them at the proper time, and see that they are comfortably sped on their way. Usually their luggage is taken down the night before, that no one may be disturbed too early in the day.

Ordinary invitations for a house-party are made to allow a part of the guests to arrive and depart at one hour, and part at another, or upon successive days, when distances to trains are far, or the number to go is large. Guests are informed when their luggage will be required for the van or wagon. Even when there are not more than one or two visitors, luggage usually precedes carriages, dog-carts or station wagons, and a trusty man looks after it properly, and sees that it is properly checked and placed upon trains, thus sparing guests all care. This is a part of a perfect hospitality that the Americans as a rule, find more difficulty, in view of their temperament, to accept than the host does to bestow. Americans usually want to be sure, by their own attention, that their belongings are safely sped ; but as the host and his attendants have thus far cared properly for them, they must submit. As a people, by and by we shall become accustomed to feeling wholly secure, while still in the hands of our hosts.

A woman bereaved of her gowns and things, and a society man who is tender of his decorative possessions, are apt to turn anxious glances at their checks as they receive them from the servants of another, and very naturally ; but they must not appear as if they doubted his discretion, or they may wound the pride of a delightful host, who has meant to be perfect in his methods for receiving and speeding his guests.

Sometimes a host drives to the station himself in a cart with one of his guests, but this is not an essential of modern hospitality. He sees that they are comfortably off from his house, in good season for their trains,

or other modes of travel, and then his part of an entertainment is completed.   If he attends them still farther,
his presence is due to friendship, courtesy, and regret
at the loss of their company.

Except when there is but one guest a hostess seldom,
if ever, goes to the train with visitors.   Even with one,
it is not a conformity with social requirements.   She
may go with guests, and so may her daughters, but it is
not expected of them.

MUTUAL OBLIGATIONS OF GUESTS AND SERVANTS.

To properly trained domestics is assigned much of the responsibility of caring for guests. To this they are trained, or at least are instructed by their employers. Their duties to visitors are clearly explained to them when they are engaged. They are to do whatever a guest desires of them when possible, and guests familiar with the duties of properly educated servants are not the ones likely to exact too much service.

As this little chapter is not written in the interest of a host's domestics, visitors who have not been in well-ordered houses, and therefore are not acquainted with the usual entertainer's customs, may be glad to know what to do, when for the first time they are guests for an extended but definite time, in a house that is thoroughly appointed with domestics.

Very soon after her arrival, the hostess may inform a woman guest that a maid is at her service at such and such times, (usually mentioning the hours before breakfast and before dinner,) to assist her in dressing, etc., and she explains perhaps that the same is her own, or her daughter's maid, or perhaps is maid for so and so of her other guests, if there are several visitors in the house. Sometimes it is the maid who offers her service to a guest, having been directed to do this by her mistress. She explains definitely what time she can give, and inquires what is likely to be wanted of her; also how the guest likes her bath to be made ready in the morning, and whether she desires coffee, etc., before dressing, etc., etc.

Many American women prefer to care for themselves, except looking after the bath, and morning cup.

A guest is very inconsiderate, and consequently illy served, if she is not ready to be attended to at the exact time which was offered as hers. If the maid assigned to her for a fixed space of time can give her no other hour, she must care for herself, because she can make no complaint to her hostess, nor can she be impatient with the maid, nor indeed with any servant in the house, since in accepting its hospitality, she has also accepted whatever her entertainer has provided for her comfort, including service that may be good or bad.

If a guest has her own maid, she is careful that this person spends her time, except by permission, in her mistress's own room, so that idling and gossip are not spread as a contagion among other servants in the house, whose duties may, or should be, more time-consuming and exacting.

Whether or not the maid is her own, for example's sake, provided her self-respect does not go so far as to influence her relations with servants, she asks them no questions about persons or things in the house, except perhaps to inquire about an invalid if there is one in the house. Should there be any lack of service in the house that is in the line of her own maid's aptitudes, while they are guests, she promptly offers her as an aid or substitute.

At less sumptuous, or perhaps only less formal and simpler houses, guests are happy, not only in taking all care of themselves, but they gladly take part with their hostess in anything she is doing that is within the limit of their attainments.

A well-educated American is as apt in serving others, as she is accomplished in the art of being elegantly served.

It is to this gift, talent, or characteristic, that she owes her adaptability to every condition of fortune, and to all varieties of giving and receiving hospitality. The maid of yesterday may be a mistress to-morrow; therefore, if for no other reason, although there is a higher one for her methods and manners with servants, a well-bred woman is always kind, considerate, and even polite to her own and to another's when they are assigned to her while a guest. Requests and not commands are given to them always. Politeness between mistress and maid is nowhere so fine and general as in France, and nowhere is service so agreeable.

When leaving, it is not only customary, but it is just, except when requested not to, to present a fee to a maid who has personally served—also to a maid of the chambers. In England every domestic who has directly or indirectly performed a service, even the lightest and most casual, expects to be remembered; but in America, only those who have been of real use are ordinarily given a fee: such as the person who carries and brings letters, and the man who takes care of the luggage; usually the groom and coachman are remembered, and sometimes a fee is sent to the cook.

These regulations apply especially to establishments where house-parties are entertained, which form of hospitality naturally creates exacting and continuous work for servants in all capacities. At houses where one, or at most, two guests are entertained at a time, the maid who is especially attentive to a guest, and perhaps a serving man, is given a fee when leaving, and this sum is in proportion to the extra amount of work a guest has caused him.

In England fees are about the same all round, except that the butler and groom get a double sum; but we have not reached that unwise usage, and many entertainers have determined that we shall not. There are hosts who ask as a favor to them, that their guests do not tip their servants, claiming that their wages have been made ample to prevent gifts from visitors.

LETTERS OF THANKS TO ENTERTAINERS.

IT is extremely discourteous to those who have entertained us to delay longer than a day after reaching home, or the house of another host, before writing letters of thanks for a hospitality that must be pronounced delightful. If guests have a happy disposition, or even if they are just, they can always find some agreeable recollections among the various experiences during any visit, and these may be, and ought to be, remembered while writing this letter. Those that are less charming should be forgotten. No matter whether the visit, in the mind of the visitor, has been an unmixed pleasure, or a somewhat diluted joy, the motive for the invitation must be held as a courtesy and friendliness ; therefore there must be prompt expressions of gratification in its remembrance.

Guests owe themselves thus much witness of gratitude, even if in the silence of their minds they regret having accepted the invitation, and have little or no satisfaction in recalling the time of their stay. It is for an intention that they must feel kindly, and for an effort to be entertaining, that they can truthfully profess to be pleased and thankful.

The graces of language are equal to this duty, and it cannot be evaded.

RECOGNITIONS OF HOSPITALITY.

IT is a common error of many persons to hold that because they cannot be as lavish or generous in return as those who invite them for extended visits propose to be, they are in honor, or perhaps delicacy, bound to decline such hospitalities.

Another mistake made by morbidly sensitive persons, is in suspecting that they are being patronized—whatever that may mean in their minds, when they are invited to visit at houses that are finer, larger or richer than their own, and they refuse, sometimes ungraciously.

It is one of the pleasantest of compensations to many persons with large wealth, who feel the burden of its care or custody, that they are able to have agreeable people under their roofs, and can give them pleasures that have been denied them by limited purses. In return, besides this satisfaction, they hope to receive much delight from minds, that, because they are not oppressed with wealth, have leisure to become rich in thought, and can foster their gifts in art, music, literature and anecdote. If accomplished people decline hospitalities that they cannot return in kind, and are sulky, or otherwise disagreeable to those who may be unfortunate in being rich, they are not noble or generous.

The kindly or noble minded Crœsus does not want returns in kind. He craves something his purse is incapable of bringing to him, and thus he depends for his best pleasures upon the graciousness of his gifted friends, or friends who are light at heart, and whose currents of interest run in other channels than his own, and such guests delight and refresh him.

A single man may send a book that is too lately issued not to be a novelty to his late hostess, and a woman who is not so placed in life that she may invite a late host to partake of hospitalities, may present a bit of her own handiwork, or give some rare trifle that has more interest than monied value in it to one who has entertained her; but even these recognitions of a much enjoyed hospitality are not obligatory and not matters of etiquette. Good Form has never meddled with them, but sometimes little gifts serve to soothe minds or hearts that feel gratitude to be a burden, and are unsatisfied with such expression of pleasure as they were able to formulate when writing their letter of thankful appreciation for delightful hospitalities.

Of course, when one who has been entertained is able to entertain in return, social life becomes a debtor and creditor affair that must, sooner or later, be balanced. This is only justice to the entertainers and their guests, and self-respect establishes a give and return order in society ; but, of course, only if circumstances are equal or nearly equal.

Not that a host who is a debtor should imitate his late entertainer in the grade or cost of a returned hospitality, but he should and can maintain the beautiful spirit of the Orientals, to whom sharing bread and salt is a sacred compact. The heart of this sentiment has never been dead, or even cold, in America, though the difficulties of establishing our civilization might have smothered it in a less kindly nation. For several centuries the entertaining of a guest was almost always a matter of duty, in which there was little or no flavor of personal pleasure or a consciousness of etiquette, but the law of kindliness was in it, and gave it sweetness and dignity.

www.ingramcontent.com/pod-product-compliance
Lightning Source LLC
Chambersburg PA
CBHW021959190326
41519CB00010B/1328